Chemical Manufacturing:
The Process of Mixing

PACT Chemical Technology Resources

Chemical Manufacturing:
The Process of Mixing

by Dianne N. Epp

Series Editor
Mickey Sarquis, Director
Center for Chemical Education

Terrific Science Press
Miami University Middletown
Middletown, Ohio

Terrific Science Press
Miami University Middletown
4200 East University Blvd.
Middletown, Ohio 45042
513/727-3269
cce@muohio.edu
© 2000 by Terrific Science Press

This monograph is intended for use by teachers, chemists, and properly supervised students. Teachers and other users must develop and follow procedures for the safe handling, use, and disposal of chemicals in accordance with local, state, federal, and institutional requirements. The cautions, warnings, and safety reminders associated with the doing of experiments and activities involving the use of chemicals and equipment contained in this publication have been compiled from sources believed to be reliable and to represent the best opinions on the subject as of 1999. However, no warranty, guarantee, or representation is made by the author or the Terrific Science Press as to the correctness or sufficiency of any information herein. Neither the author nor the publisher assumes any responsibility or liability for the use of the information herein, nor can it be assumed that all necessary warnings and precautionary measures are contained in this publication. Other or additional information or measures may be required or desirable because of particular or exceptional conditions or circumstances, or because of new or changed legislation.

ISBN: 1-883822-23-8

This material is based upon work supported by the National Science Foundation under Grant Number DUE 9751993. Any opinions, findings, and conclusions or recommendations expressed in this material are those of the author and do not necessarily reflect the views of the National Science Foundation.

Table of Contents

● Acknowledgments

The author and editor wish to thank the following individuals who have contributed to the development of *Chemical Manufacturing: The Process of Mixing:*

Terrific Science Press Design and Production Team

Document Production Manager: Susan Gertz
Technical Editing: Lisa Taylor, Kim Jacobs
Illustrations: Carole Katz, Becky Franklin, Brian Fair
Design/Layout: Susan Gertz, Lisa Taylor
Production: Lisa Taylor, Jenny Stencil, Becky Franklin, Lisa Alexander, Brian Fair, Kim Jacobs, Jan Garcia, Tom Schaffner
Cover Design: Susan Gertz

Reviewers and Classroom Testers

Paul W. Barnett, Whirlpool Corporation, Benton Harbor, MI
William Bleam, Jr., Radnor High School, Radnor, PA
Nancy LeMaster, D.W. Daniel High School, Central, SC
E. Bruce Nauman, Rensselaer Polytechnic Institute, Troy, NY
Amy Stander, Center for Chemical Education, Middletown, OH
Melanie Stewart, Stow-Munroe Falls High School, Stow, OH
Linda Woodward, Miami University Middletown, Middletown, OH

To the many fine students I have met along the way; you are why I teach, so this one's for you!

Foreword

Chemical Manufacturing: The Process of Mixing is a comprehensive instructional unit for high school chemistry. It provides teachers with complete lesson plans, teacher background, student handouts, overheads, student labs, and a student research project. You may copy these materials as necessary for classroom use.

While numerous unit operations can be investigated in the chemical manufacturing process, mixing is one of the most common processes conducted in industry. This monograph investigates this fundamental process as an engineering problem of achieving a satisfactory mixture in a given industrial situation and also looks at the effect mixing has upon the quality of the final product. The lessons expose students to the reasoning process involved in solving mixing problems, to practical examples of the application of chemistry and chemical engineering in industry, and to insights into what chemical engineers and technicians do.

This monograph is part of the PACT Chemical Technology Resources series, which is a product of the National Science Foundation-funded Partnership for the Advancement of Chemical Technology (PACT) program at Miami University Middletown in Ohio. PACT is an industrial/academic collaborative committed to creating a well-educated, chemistry-based technical workforce. Members of the PACT Consortium share the goal of bringing chemistry and chemical technology education into closer alignment with the skills, methods, problem-solving, and content used in today's industrial and government laboratories.

Author Dianne Epp, a teacher at East High School in Lincoln, Nebraska, has more than 20 years of teaching experience as well as industrial research experience. She is a highly respected author of resource books for high school chemistry teachers. In this volume Dianne brings together the science and engineering of mixing in the chemical manufacturing industry with innovative classroom activities.

We hope you will find this monograph to be a useful and exciting tool for involving your students in workplace applications of chemistry. You can find out more about the PACT Consortium and Chemical Technology Resources series by accessing our Web site at *www.terrificscience.org/PACT/*.

Mickey Sarquis, Director
Center for Chemical Education
June 1999

● Preface

The purpose of this monograph is to provide students with a window into the
world of the chemical engineer. The chemical engineer is familiar with the
applications of chemistry in large-scale production processes. One production
process that is important in all manufacturing operations is the appropriate
mixing of ingredients. Since this process is common to all industrial operations,
from the mixing of raw ingredients for plastics and pharmaceuticals to the
blending of materials in a mixed product such as a solid fertilizer mix or a blend
of grass seed, it is a process requiring the attention of the chemical engineer. By
carrying out the activities in this monograph students investigate the problems
that arise in preparing a well-mixed sample and then investigate the importance
of adequate mixing to the quality of the final product.

Curriculum Placement

This monograph may be used as part of an introduction to careers in science, as
part of a unit on the physical properties of matter, or as a link to technology and
historical perspectives in the chemistry curriculum.

Relationship to National Science Education Standards

The National Science Education Standards prepared by the National Research
Council are supported by the use of this material. Overall, this monograph is
laboratory inquiry-based, which reinforces the concept of science as inquiry.
Also, the use of industrial processes and techniques helps students understand
the connection between science and technology.

Section 1, "Where Are the Raisins?," challenges students to hypothesize, design
and conduct an experiment to test their hypotheses, analyze their results, and
draw logical conclusions concerning those results. This process enables students
to continue to develop skills in science as inquiry, which addresses Content
Standard A, Science as Inquiry. Section 1 also contains a significant amount of
historical information that may be used to develop student understanding of
science as a human endeavor, as outlined in Content Standard G, History and
Nature of Science.

Section 2, "Understanding the Mixing Problem," also contains historical
information, which supports Content Standard G, History and Nature of Science.
The discussion of various types of industrial mixers in this section ties science to
technology, as outlined in Content Standard E, Science and Technology.

Section 3, "Why Do Particles Segregate?," requires an understanding of the structure and properties of matter and motions and forces as outlined in Content Standard B, Physical Science. In order to understand segregation patterns illustrated in various laboratory activities, the challenge activity in this section may be used as an authentic assessment tool to determine whether students have grasped the concept of segregation as a function of particle size.

Section 4, "Where Does the Bounce Come From?," is based on crosslinked polymer chemistry and addresses the chemical reaction section of Content Standard B, Physical Science. Again, as a laboratory inquiry it allows students to develop expertise in scientific inquiry, designing experiments, analyzing data, assessing results and communicating information.

Section 5, the culminating research activity, again reinforces Content Standard E, Science and Technology, by engaging students in a design task to formulate a particular product within the constraints of available materials.

Cross-Curricular Integration

The following are suggestions for activities to integrate the study of mixing as a chemical engineering topic with other disciplines.

Food Science

- Many examples of mixing are present in food science, especially in baking processes. Have students set up controlled experiments to test the effect of various amounts of mixing on such products as muffins and biscuits. What happens to the quality of the product if the ingredients are insufficiently mixed? Is it possible to "overmix" in some cases? Research the nature of doughs and batters with respect to mixing phenomena. This issue, which is of obvious importance to the home baker, is also a major area of industrial research.

Economics

- Have students research and report on various ways in which a manufacturing operation may be made more economically efficient. Issues to be addressed may include equipment size, automation, and continuous processes versus batch processes.

History

- Have students research the history of a particular chemical manufacturer, such as DuPont, Bayer, Corning Glass, Kodak, or BASF, or the development of a particular chemical process such as the vulcanization of rubber or the cracking of petroleum.

Safety Procedures

Experiments, demonstrations, and hands-on activities add relevance, fun, and excitement to science education at any level. However, even the simplest activity can become dangerous when the proper safety precautions are ignored or when the activity is done incorrectly or performed by students without proper supervision. While the activities in this book include cautions, warnings, and safety reminders from sources believed to be reliable, and while the text has been extensively reviewed, it is your responsibility to develop and follow procedures for the safe execution of any activity you choose to do. You are also responsible for the safe handling, use, and disposal of chemicals in accordance with local and state regulations and requirements.

Safety First

- Collect and read the Materials Safety Data Sheets (MSDS) for all of the chemicals used in your experiments. MSDS's provide physical property data, toxicity information, and handling and disposal specifications for chemicals. They can be obtained upon request from manufacturers and distributors of these chemicals. In fact, MSDS's are often shipped with the chemicals when they are ordered. These should be collected and made available to students, faculty, or parents for information about specific chemicals used in these activities.

- Read and follow the American Chemical Society Minimum Safety Guidelines for Chemical Demonstrations on the next page. Remember that you are a role model for your students—your attention to safety will help them develop good safety habits while assuring that everyone has fun with these activities.

- Read each activity carefully and observe all safety precautions and disposal procedures. Determine and follow all local and state regulations and requirements.

- Never attempt an activity if you are unfamiliar or uncomfortable with the procedures or materials involved. Consult a college or industrial chemist for advice or ask him or her to perform the activity for your class. These people are often delighted to help.

- Always practice activities yourself before using them with your class. This is the only way to become thoroughly familiar with an activity, and familiarity will help prevent potentially hazardous (or merely embarrassing) mishaps. In addition, you may find variations that will make the activity more meaningful to your students.

- You, your assistants, and any students participating in the preparation for or doing of the activity must wear safety goggles if indicated in the activity and at any other time you deem necessary.

- Special safety instructions are not given for everyday classroom materials being used in a typical manner. Use common sense when working with hot, sharp, or breakable objects. Keep tables or desks covered to avoid stains. Keep spills cleaned up to avoid falls.

American Chemical Society Minimum Safety Guidelines for Chemical Demonstrations

This section outlines safety procedures that Chemical Demonstrators must follow at all times.

1. Know the properties of the chemicals and the chemical reactions involved in all demonstrations presented.

2. Comply with all local rules and regulations.

3. Wear appropriate eye protection for all chemical demonstrations.

4. Warn the members of the audience to cover their ears whenever a loud noise is anticipated.

5. Plan the demonstration so that harmful quantities of noxious gases (e.g., NO_2, SO_2, H_2S) do not enter the local air supply.

6. Provide safety shield protection wherever there is the slightest possibility that a container, its fragments, or its contents could be propelled with sufficient force to cause personal injury.

7. Arrange to have a fire extinguisher at hand whenever the slightest possibility for fire exists.

8. Do not taste or encourage spectators to taste any non-food substance.

9. Never use demonstrations in which parts of the human body are placed in danger (such as placing dry ice in the mouth or dipping hands into liquid nitrogen).

10. Do not use "open" containers of volatile, toxic substances (e.g., benzene, CCl_4, CS_2, formaldehyde) without adequate ventilation as provided by fume hoods.

11. Provide written procedure, hazard, and disposal information for each demonstration whenever the audience is encouraged to repeat the demonstration.

⓬ Arrange for appropriate waste containers for and subsequent disposal of materials harmful to the environment.

Revised 4/1/95. Copyright© 1995, ACS Division of Chemical Education, Inc. Permission is hereby granted to reprint or copy these guidelines providing that no changes are made and they are reproduced in their entirety.

Teacher Background

Chemical Engineering

Chemical engineers use their knowledge of mathematics, science (especially chemistry), and engineering to overcome technical problems safely and economically. Chemical engineers came into demand in the Industrial Revolution when a skill set that bridged the traditional domains of chemists and engineers became necessary. While chemists typically lacked the knowledge needed to scale up a laboratory experiment to industrial-sized batches, mechanical engineers did not have the knowledge of chemistry that was needed to make many large-scale processes possible. Chemical engineers were familiar with both chemistry and production.

Here are some specific examples of work that chemical engineers might do:

- improve food processing techniques and methods of producing fertilizers in order to increase the quantity and quality of available food;

- construct synthetic fibers that make our clothes more comfortable and water-resistant;

- develop methods to mass-produce drugs, making them more affordable;

- create safer, more efficient methods of refining petroleum products, making energy and chemical sources more productive and cost-effective; and

- develop solutions to environmental challenges, such as pollution control and remediation.

Chemical engineers can also work in fields such as law, education, publishing, finance, and medicine.

Much of this information came from the web site for the American Institute of Chemical Engineers (*www.aiche.org/*). Your students can find additional information, such as salary statistics, at that web site.

Mixing

Mixing is the intermingling of several dissimilar materials in an attempt to produce a desired level of uniformity. Chemical manufacturing, as well as other industries that use chemical and physical changes in their manufacturing processes, invariably deal with mixing operations.

Much of the knowledge concerning mixing operations has resulted from work in the chemical industry, although many other manufacturing industries also deal with mixing operations on a routine basis. The ability to mix diverse materials to form a homogenous batch in a manufacturing operation is of primary importance, because the homogeneity of the batch going into the process and the homogeneity of the material as it goes through the manufacturing process relate directly to the uniformity (and therefore quality) of the product.

The chemical engineer may be faced with a variety of mixing problems, largely based on the states of matter of the components. Blending miscible liquids, such as the blending of petroleum products, can be relatively simple. In other cases, such as emulsification processes often encountered in the food and pharmaceutical industries, mixing presents a greater challenge, because the components are in different states of matter. Physical and chemical changes in a production process may also involve mixing problems. This monograph introduces students to some of the problems a chemical engineer might encounter in mixing processes.

The goal of the mixing process is to achieve a fully randomized sample. Components of any mixture consist of particles capable of relative movement, but the particles in a solid have no translational motion without external forces. Therefore, in a solid mixture, the rate at which mixing occurs depends on the movement of the particles caused by external handling or other forces. In this way, solids are unlike liquids and gases, in which molecular diffusion plays a significant role in mixing. Figure 1 illustrates three possible configurations of dissimilar solid particles. Note that even in sample b, which is considered fully randomized, the situation is not perfect; several white particles are surrounded mostly by other white particles. Overall, however, the sample must be considered completely mixed, because a sample from any part of the batch would contain a representative composition, and further mixing would not reduce local fluctuations.

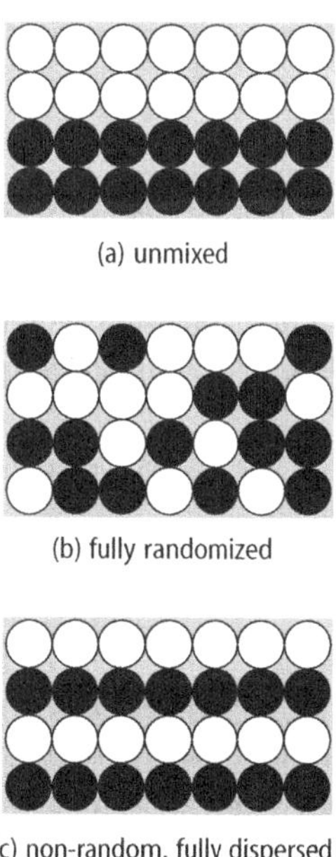

(a) unmixed

(b) fully randomized

(c) non-random, fully dispersed

Figure 1: Particle distribution in mixtures

Particles in a solid mixture may differ widely in physical characteristics. In manufacturing processes involving solid materials, these differences must be given careful consideration. The term "segregation" describes the result of the natural tendency of a solid mixture to "unmix" due to differences in (in order of importance) the particles' size, density, and shape. Segregation, or the variation in composition from point to point, occurs within a sample when differences in properties cause some particles to move selectively relative to other particles.

As noted previously, the size differential of solid particles within a sample is the most important factor in segregation. Even relatively small differences in particle size are enough to allow significant segregation, because smaller particles can easily slip and slide past the larger particles.

One method of controlling segregation in a mixture is to reduce the particle size of all components. When the size of all particles in the solid is reduced by grinding, segregation decreases, because bonding forces between the particles, such as van der Waals forces, electrostatic attractions, and surface moisture on the particles become significant factors. Research has shown that when the particle size is below 10 micrometers in diameter, no appreciable segregation occurs.

Particle shape may also play a significant role in mixing. If the major component has a rough or fibrous shape, very little segregation may occur. If one of the components is extremely fine, it may "coat" the larger particles. In this case, the very fine particles lose their individual freedom of movement, and the sample mixes very well.

The most obvious way to ensure homogeneity is to grind all components to a similar particle size. However, since exact matching of particle size is difficult at best and sometimes impossible due to the nature of the components, grinding will generally minimize rather than eliminate segregation.

Any mixture, if looked at closely enough, will show areas of segregation. The size of the regions of segregation that can be tolerated in a manufacturing process varies from case to case. Whether a sample is satisfactorily mixed depends on the end use of the product. For example, the blending of pigment to give an apparent uniform color depends on the resolving power of the eye and the distance from which the pigment is observed. At a distance of 1 foot, segregated areas may be clearly visible, but at a distance of 10 feet they may blend to appear uniform, like the dots of color in a pointillist painting. In a mixed product such as a fertilizer blend, the overall composition of one bag must be the same as that in another bag. When raw materials are mixed to produce a product, such as when sand and soda ash are mixed to make glass, the acceptable level of segregation in the solid glass batch will be determined by the composition variance acceptable in the finished product.

References

American Institute of Chemical Engineers (AIChE) Home Page. http://www.aiche.org/ (accessed Oct 1999).

Chain Gang—The Chemistry of Polymers; Sarquis, M., Ed.; Science in Our World Series; Terrific Science: Middletown, OH, 1995.

Chemical Engineers' Handbook; Perry, J.H., Ed.; McGraw-Hill: New York, 1963.

The Corning Glass Center; Corning Glass Works: Corning, New York, 1968.

Danckwerts, P.V. *Insights Into Chemical Engineering;* Pergamon: Oxford, 1981.

Encyclopedia of Chemical Technology; Kirk, R., Orthmer, D. Eds.; Interscience: New York, 1972.

Griskey, R.G. *Chemical Engineering for Chemists;* American Chemical Society: Washington, DC, 1997.

Hobby, G.L. *Penicillin: Meeting the Challenge;* Yale University: New Haven, CT, 1985.

Killeffer, D.H. *Chemical Engineering;* Doubleday: New York, 1967.

McCabe, W.; Smith, J. *Unit Operations of Chemical Engineering;* McGraw-Hill: New York, 1967.

McDonough, R.J. *Mixing for the Process Industries,* Van Nostrand Rheinhold: New York, 1992.

Mixing in the Process Industries; Harnby, N., Edwards, M.F., Nienow, A.W., Eds.; Butterworth: London, 1985.

Morton, M. *Rubber Technology;* Van Nostrand Reinhold: New York. 1987.

Orr, C. *Particulate Technology;* Macmillan: New York, 1966.

Ratcliff, J.D. *Yellow Magic;* Random House: New York, 1945.

This Is Glass; Corning Glass Works: Corning, New York, 1957.

Uhlmann, D.R., Kreidl, N.J. *Glass: Science and Technology;* Academic: Orlando, FL, 1984; Vol. 2.

Whitehead, D. *The Dow Story;* McGraw-Hill: New York, 1968.

Section 1
Where Are the Raisins?

● An Analysis of Raisin Bran (Student Activity 1)

Does your box of raisin bran contain "two scoops" of raisins? When you open a new box of raisin bran are you more likely to find the raisins at the top or at the bottom? And when you pour out your morning bowl of cereal, how many raisins are you likely to find if you are getting near the bottom of the package? These questions deal with the homogeneity of a sample that is composed of particles of different shapes, sizes and densities.

Safety and Disposal

Although a food sample is being used in this activity, do not eat any of it. No special disposal procedures are required.

Materials

Per group

- box of raisin bran (any brand)
- 3 large, flat pans
- felt-tipped marker
- masking tape
- ruler
- 250-mL beaker, scoop, or cup

Procedure

❶ As a group, discuss the following questions and record your predictions:

- Will there be more raisins in the top third of the box or the bottom third?

____

____

- Will there be any difference in the number of raisins in each third of the box?

____

____

- Will turning the box over a number of times to mix the contents have any noticeable effect on the distribution of raisins in the box?

____

____

❷ Use the masking tape and marker to label the three pans "top third," "middle third," and "bottom third."

❸ Do one of the following steps, depending on your group's assigned number.

- Groups 1 and 2: Carefully remove the inner waxed bag containing the cereal and measure the approximate height of the cereal. Divide the cereal into equal thirds, by height, marking the thirds on the outside of the bag with a felt-tipped marker.

- Groups 3 and 4: Before opening the box, rotate it end over end 100 times to mix the contents of the box. (See Figure 1-1.)

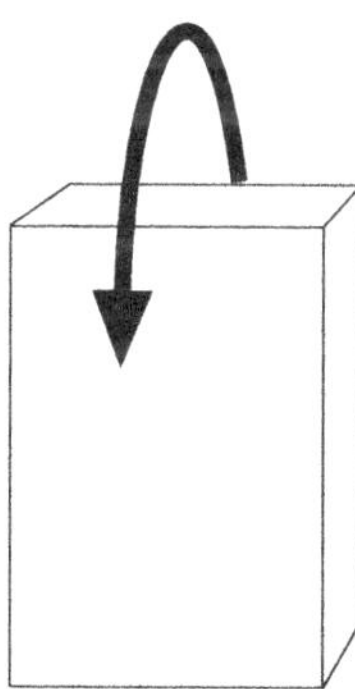

Figure 1-1: Rotate the box end over end.

Carefully remove the inner waxed bag containing the cereal and measure the approximate height of the cereal. Divide the cereal into equal thirds, by height, marking the thirds on the outside of the bag with a felt-tipped marker.

- Groups 5 and 6: Before opening the box, rotate it horizontally 100 times to mix the contents of the box. (See Figure 1-2.)

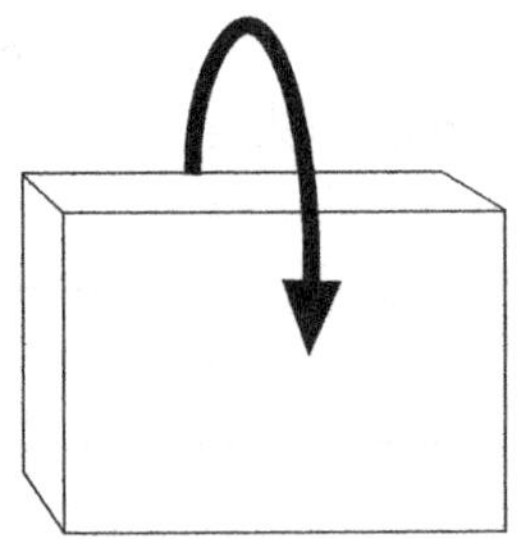

Figure 1-2: Rotate the box horizontally.

Carefully remove the inner waxed bag containing the cereal and measure the approximate height of the cereal. Divide the cereal into equal thirds, by height, marking the thirds on the outside of the bag with a felt-tipped marker.

4 Open the bag carefully and use the 250-mL beaker, scoop, or cup to scoop out the top third of the cereal, placing it in the large, flat pan labeled "top third."

5 Repeat step 4 two more times, placing the middle third of the cereal in the second labeled pan and the bottom third in the third labeled pan.

6 Count the number of raisins and calculate the percentage found in each sample. Record your data and compare results with those of the other groups. Discuss the results in light of your predictions and discuss characteristics of the bran flakes and the raisins that may have contributed to the observed results.

____

____

____

____

____

Extension

Consider the following case: A can of mixed nuts typically contains peanuts, Brazil nuts, cashews, almonds, and pecans. (See Figure 1-3.)

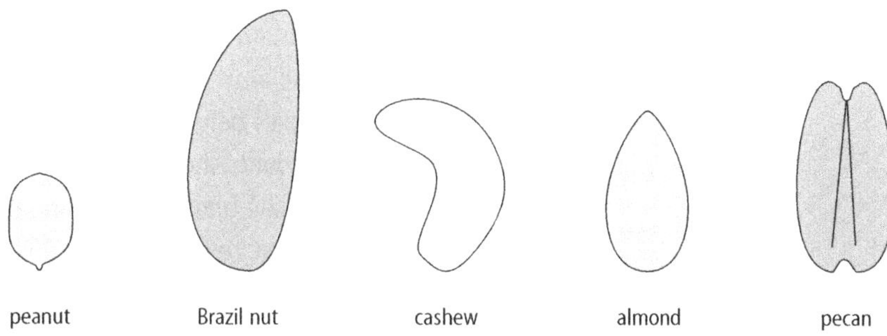

Figure 1-3: A can of mixed nuts typically contains five kinds of nuts.

In a typical can of mixed nuts, will the five types be evenly distributed throughout the can? Record your response along with your reasoning. If you do not think the five types will be evenly distributed, draw a generalized distribution of what you would expect to find, by type, in the bottom half of the can versus the top half of the can.

____

____

____

____

This investigation is designed to be done with six groups of students. Two groups carry out each analysis procedure, and then all groups pool information by posting it on the chalkboard or a large class chart. Two groups will analyze the raisin bran as the boxes arrive from the store, and the other four groups will mix the contents of the boxes before analysis by rotating them vertically or horizontally 100 times. Students should be cautioned not to shake the boxes, because shaking will crush the bran flakes, changing the size of the flakes and therefore changing the overall result. As an extension after the activity is finished, students might thoroughly crush the bran flakes, return the crushed flakes and raisins to the boxes, and then shake or rotate the boxes to determine whether the raisin distribution changes.

The Extension with the mixed nuts may be carried out in groups, as with the raisin bran investigation, or it could be conducted solely as a prediction exercise based on the findings from the raisin bran investigation.

Safety

Remind students never to eat anything in the laboratory. No special disposal procedures are required.

Sample Data

In a box that has not been rotated, a typical distribution from our laboratory testing was as follows (the actual numbers of raisins will vary with the brand analyzed, but the relative distribution will be approximately the same):
- upper third = 30 raisins (12%)
- middle third = 80 raisins (34%)
- bottom third = 128 raisins (54%)

Raisins segregate to the bottom part of the container as the smaller, more dense raisins move through the spaces between the larger, less dense flakes.

In a box that has been rotated vertically 100 times, the following distribution was typical in our lab testing:
- upper third = 92 raisins (36%)
- middle third = 74 raisins (28%)
- bottom third = 92 raisins (36%)

While the distribution was more equal than with no rotation, the raisins appear to segregate somewhat to the top and bottom of the box.

In a box that has been rotated horizontally 100 times, the following distribution was typical in our lab testing:
- upper third = 75 raisins (30%)
- middle third = 85 raisins (33%)
- bottom third = 95 raisins (37%)

Horizontally rotating the box seems to mix the raisins into the three regions, although higher numbers still occur in the lower portions of the box.

Regarding the Extension, some brands of mixed nuts contain filberts as well as the other five nuts listed in the activity. In a typical can, a higher proportion of larger nuts is likely to be found near the top, while the smaller nuts will tend to have settled toward the bottom. (See Figure 1-4.) Other foodstuffs that exhibit segregation include dry soup mixes or packets of hot chocolate mix containing miniature marshmallows. In the latter case, the wide difference in size between the marshmallows and the hot-chocolate mix emphasizes how the material with smaller particle size sifts between the larger particles.

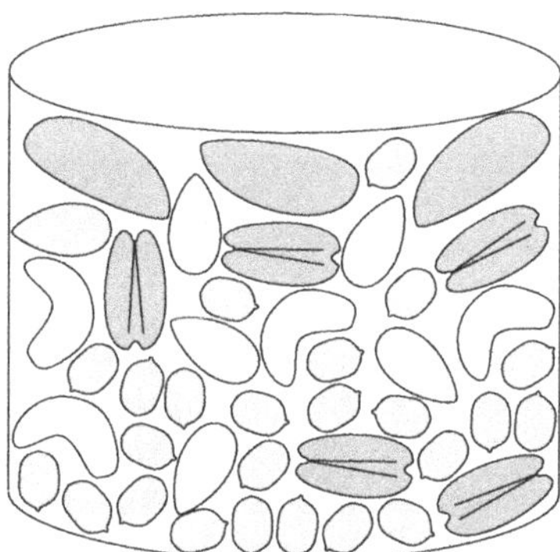

Figure 1-4: In a typical can of mixed nuts, a higher percentage of larger nuts is likely to be found near the top, while the smaller nuts will tend to have settled toward the botom.

Section 2

Understanding the Mixing Problem

● Peas and Honey: The Importance of Mixing (Student Background 2A)

I eat my peas with honey;
I've done it all my life,
They do taste kinda funny,
But it keeps 'em on the knife.

—Ogden Nash

While the question of mixing such unusual items as peas and honey is not one encountered on a daily basis, those who work in the chemical process industry, manufacturing everything from rubber tires for automobiles to mayonnaise for sandwiches, do face the problem of mixing materials that may be just as different from one another as those mentioned in the rhyme above. And just as above, the reason for mixing them may have to do with their unique properties and the way they interact with their surroundings, as the following two examples illustrate.

In 1928, Sir Alexander Fleming observed the death of bacteria in a Petri culture dish when certain types of dust particles dropped into the culture. Experimentation over a number of years led him to conclude that the molds of the common *Penicillium* genus, often found on moldy bread and cheese, produce a substance that is deadly to certain types of disease-causing bacteria. The first clinical trials of the new antibiotic were carried out in Great Britain in 1941, when the need to treat World War II battle wounds gave the work a particular urgency. When the antibiotic was proven particularly effective, producing a large amount of the mold presented problems. In order to grow, the mold required plenty of food and plenty of air. The mold could receive nutrients from the surface of a concentrated corn syrup, but as the mold grew it formed a heavy blanket. This blanket was problematic because on the underside, the mold had plenty of food but no air, while on the upper side the mold had plenty of air but no food. Fortunately, engineers found that bubbling sterilized air through the nutrient solution solved the problem. The air bubbles provided the needed air supply to the underside of the mold, and the stirring action of the bubbles kept all parts of the mold in contact with the nutrient solution. The need for materials to be in contact in order to react is only one of the reasons why mixing is a major industrial issue.

Another example of a process in which mixing is important is glassmaking. The making of glass often requires mixing batches of four or more separate materials having different mixing characteristics. The glass batch typically includes sand (SiO_2), soda ash (Na_2CO_3), limestone ($CaCO_3$), and a certain quantity of cullet (crushed glass of the same type that is being produced). Such mixtures are inherently difficult to mix well, but a large body of experience shows that care taken during batch mixing considerably improves the quality of the glass. Modern glass furnaces are giant tanks holding up to hundreds of tons of glass, but the beginning of any successful manufacture is in the batch house, where the raw ingredients are mixed before being fed into the furnace. The problem of keeping the batch well mixed as it is transported and fed into the furnace presents the glassmaking team with a significant challenge. The nature of the ingredients in the batch, such as their particle size and shape, density, and ability to flow, all affect the mixing of the solid ingredients in the batch. In a glass batch the characteristics of the raw materials differ significantly, and if the batch segregates significantly—that is, if certain chemicals tend to clump together—the quality of the glass will be uneven as the glass progresses through the furnace and is converted into a viscous mass. This unevenness will show up as bubbles or other imperfections in the final formed product. Any errors in the mixing of the raw materials will assuredly result in a glass quality problem. Thus, it is critical that a carefully mixed solid batch be fed into one end of the furnace so that the molten mixture flowing into the forming machines (see Figure 2-1) at the lower end of the furnace can be of the highest quality possible. In glass production, as in many industrial processes, the quality of the end product is the final test of mixing.

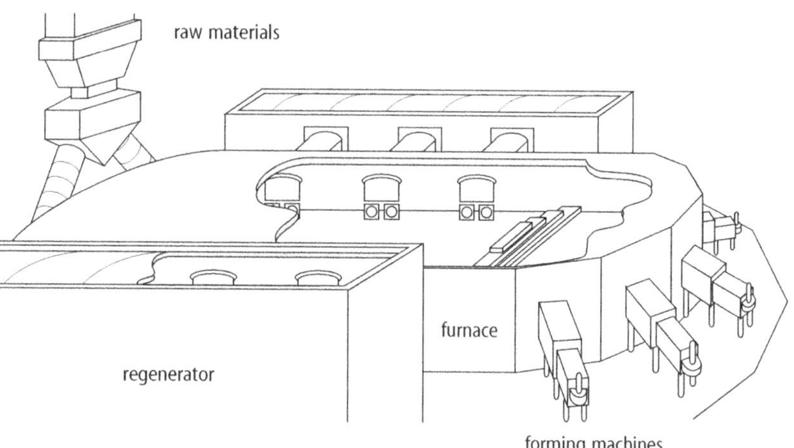

Figure 2-1: A typical glass furnace

Mixing is a central feature of processes in many industries, including the food, pharmaceutical, paper, plastic, and rubber industries, to cite just a few. Therefore, chemical engineers must have a good understanding of the principles of mixing.

When a new medicine is discovered in the laboratory it may create much excitement among the researchers, but if it doesn't move out of the laboratory it is of little value. It can't cure diseases or save lives until a manufacturing company starts producing the material in large quantities. Only then can it find its way into hospitals and into the hands of physicians. Both researchers and industrialists must work together to bring a new drug into use. The story of the development of penicillin dramatically illustrates the importance of this partnership.

Act One: Alexander Fleming (see Figure 2-2), born in Scotland in 1881, studied medicine in London. In the fall of 1928 he was teaching and conducting research at St. Mary's in London. His research centered on cultures of staphylococci, the microbes that cause boils, blood poisoning, and a host of other ills. His cultures of these bacteria filled dozens of Petri dishes. One fall day, Fleming noticed that one of the culture plates had become contaminated with a bit of mold, not an uncommon occurrence generally resulting in the culture's being thrown away. Yet on that day he hesitated, for his experienced eye and prepared mind saw something that other bacteriologists might have missed. His lab notebook records, "It was astonishing that for some considerable distance around the mold growth the staphylococcal colonies were undergoing lysis." The term "lysis" means that something was splitting the colonies apart and causing them to disintegrate.

Figure 2-2: A portrait of Alexander Fleming

Fleming and two laboratory helpers, Ridley and Craddack, began testing the medium in which the mold was growing and found that even at dilutions of one part to 800 (1:800), the solution was effective in killing the staphylococci. It proved equally effective against streptococci, the microbes that cause pneumonia. When animal studies showed the material to be nontoxic, Fleming christened it penicillin, as it was derived from a mold belonging to the *Penicillium* genus (see Figure 2-3), and presented his findings in a 1929 journal article. In spite of its amazing ability to destroy microbes, a major hurdle remained: the mold filtrate contained only minute amounts of the drug. Gallons of the material would be needed to control even minor infections. Clearly, there was work yet to be done.

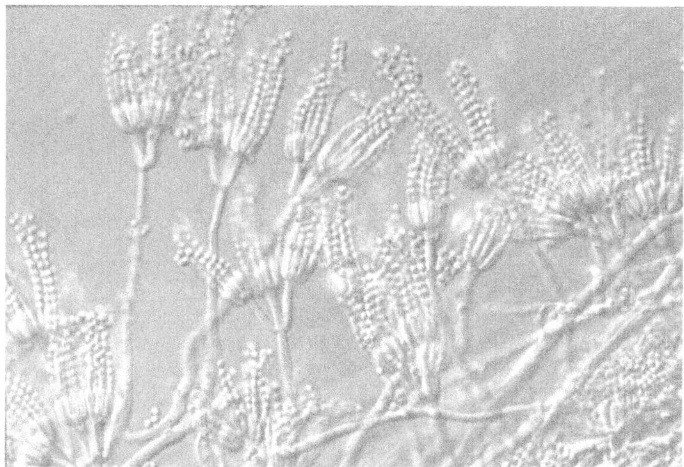

Figure 2-3: A micrograph showing rows of spores common in Penicillium

Act Two: While Fleming had established the antibacterial properties of penicillin, the practice of treating diseases by using chemicals to kill bacteria was still in its infancy. In 1910, Paul Erlich had opened the era with his discovery of the arsenic compound salversan, which killed the syphilis microbe, but it was not until the discovery of sulfa drugs in the early 1930s that research in using chemicals to cure disease gained wide attention. Some researchers had worked briefly with "Fleming's mold" but found the penicillin extremely difficult to extract from the culture. Because of this difficulty, plus the fact that penicillin is exceedingly perishable, many attempts to isolate the substance ended in failure.

In 1938 a new player came on the scene. Australian-born Dr. Howard Florey, a researcher at Oxford University in England, had gathered a research group to study bacterial antagonism—the toxic effect of one bacterium on another. A literature survey by his associate, Dr. Ernst Chain, included the original article by

Fleming on the antibacterial qualities of penicillin, and Florey and his research team proceeded to grow quantities of the mold and isolate the penicillin from the broth in which it was formed. During the next several years Florey's research team (see Figure 2-4) painstakingly isolated and experimented with penicillin, and some of their clinical trials showed penicillin effective where sulfa drugs had failed. However, the mountain of labor required to isolate enough of the drug to treat even one patient remained a significant obstacle. It was time for the process to leave the laboratory and enter the industrial world.

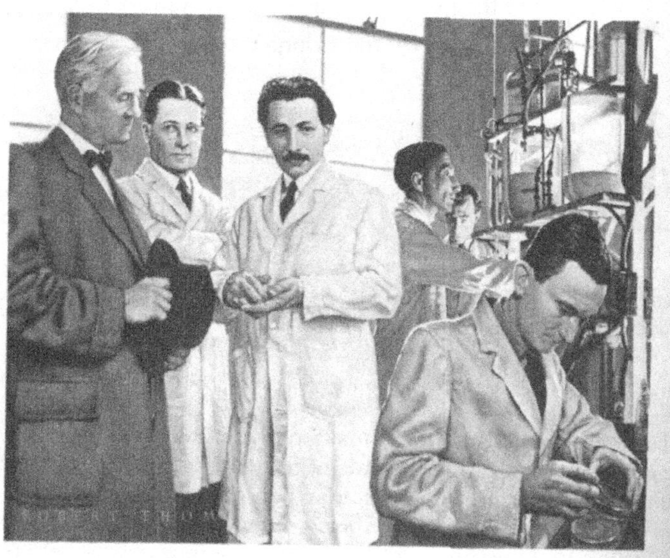

Figure 2-4: In 1940, Alexander Fleming (far left) met with the research team who had reisolated penicillin and were testing it on mice at the Sir William Dunn School of Pathology, Oxford, England. This painting by Robert Thom shows how this meeting might have looked. The team was led by Howard Florey and Ernst Boris Chain (standing, second and third from left). Norman Heatley, another member of the research group, is seated at right. Fleming, Florey, and Chain shared the Nobel Prize five years later.

Act Three: In June of 1941 British drug companies were interested in producing penicillin, but their resources were entirely committed to the war effort. Florey, convinced that penicillin was vital in treating war wounds, traveled to the United States to seek manufacturing assistance. Meetings with Dr. Robert Coghill of the U.S. Department of Agriculture provided the impetus for an all-out industrial assault on the problems surrounding the production of penicillin. In an unprecedented collaborative effort between the government and pharmaceutical firms such as Merck, Squibb, Pfizer, and Lederle, massive resources were poured into the production effort. One hurdle, that of finding the most favorable culture for the growth of the mold, was overcome when corn-steep liquor (the fluid left after corn is soaked in water in the starch-making industry) was added to the culture. The second major breakthrough came when it was found that the mold could be grown in a submerged culture in large fermentation vats if a continuous stream of air was bubbled into the fermenting vats. The mixing of the air into the solution by the bubblers allowed mold, food, and oxygen to be in constant contact. Without this critical mixing, the mold would die. The amounts of penicillin produced by Merck, Squibb, Pfizer, and others in the industry increased steadily from 1941–1943, and by 1945 the supply of penicillin was rising faster than the demand. Thus penicillin moved from a tiny spot observed on Fleming's Petri dish to a major weapon in the battle against disease.

● History of the Glass Industry (Student Background 2C)

No written record tells us when the first glass objects were manufactured, but
the art of glassmaking is at least 3,500 years old. By 1500 B.C. glass beads and
small glass containers were being produced. Fascinatingly, it is only in the last
several hundred years that significant changes have been made in the basic
glass batch: silica in the form of sand, alkali from soda ash, and lime. Until the
17th century, the only real improvement in this soda-lime glass was in the
purification of materials. Then in 1676, an Englishman named George
Ravenscroft obtained a glass of unusual clarity by adding lead oxide to the glass
batch. This addition also produced a much softer glass that was much easier to
cut and engrave, and the era of lead crystal was born. In the late 18th century, a
Swiss glassmaker, Pierre Louis Guinand, developed a method of stirring the
molten glass. This mixing technique produced a more homogeneous glass with
fewer streaks from unmelted materials. Overhead 2-1 shows an illustration of an
early glassworks, as it appeared in Diderot's *L'Encyclopédie, ou Dictionnaire
Raisonné des Sciences, des Arts et des Métiers,* 1751.

Expanding the types of glass available by changing the basic batch recipe began
seriously in the 19th century. Instrumental in this development was the German
industrialist Otto Schott, who, along with his associates, introduced more than
25 new ingredients for use in glassmaking. Colored glasses resulted from the
addition of small amounts of metal oxides to the glass batch: cobalt with copper
for blues; iron for amber; and gold, copper or selenium for reds.

The first successful glassworks in the United States was founded in 1739 by
Caspar Wistar in Wistarberg, NJ. Mass production of glass came about in the
early 1800s when American engineers perfected a method of pressing hot glass
into a metal mold. The ability to produce inexpensive glass bottles
revolutionized the container industry.

During the 20th century the glass industry has greatly expanded the variety of
glass formulations so that tens of thousands of workable glass compositions
now exist that make use of almost every element on the Earth's surface. Corning

Glass Works in Corning, NY, began the expansion in 1912 with the formulation of heat- and chemical-resistant borosilicate glasses. In succeeding years Corning alone has recorded the properties of more than 100,000 glassmaking formulas and each week melts approximately 200 new experimental formulations.

American glassmakers annually produce about 8 million tons of glass worth approximately 2 billion dollars. Automated machines allow container companies such as Owens-Illinois Glass Company to turn out millions of glass items annually. Flat glass for industrial and residential buildings as well as for use in automobiles is produced by companies such as Pittsburgh Plate Glass and Libbey-Owens-Ford Glass. Corning Glass Works has been the leader in specialty glasses, manufacturing thousands of products with specific, and often unique, properties. The ability of all of these companies to turn out a quality product is directly related to their ability to deliver a well-mixed glass batch to the furnace. If a glass batch is not well mixed, the entire melting process, as well as the final quality of the glass as it reaches the forming machines, will be adversely affected.

● Overhead 2-1: Historical Illustration of Glassmaking

An eighteenth-century glassworks (from A Diderot Pictorial Encyclopedia of Trades and Industry, a collection of plates from Diderot's L'Encyclopédie, ou Dictionnaire Raisonné des Sciences, des Arts et des Métiers)*

*This picture filled two facing pages in the book it was scanned from. The dark line down the center of the picture is the crease of the book, not part of the picture.

Who Solves the Problems? The Chemical Engineer! (Student Background 2D)

The "good old days" perhaps were not so good as writers and artists might portray them. Modern airplanes would not leave the ground if powered with the fuels of the 1920s, and automobile tires of the same era lasted only 3,000–5,000 miles at best compared to 25,000–30,000 or higher today. (Of course, at 25 miles per hour cars did not cover a lot of distance in a short period of time.) From the myriad of rich colors that dye our fabrics to the great variety of the fabrics themselves, from the television and films that entertain us to the medicines we use to fight many formerly fatal illnesses, we enjoy many things that did not exist in the "good old days" of our great-grandparents.

None of these things "just happened." Technological progress often involves cooperative efforts among various teams in industry, including research or development scientists, engineers, and production and quality control teams supported by laboratory and process technicians. Specifically, chemical engineers convert the abstract, theoretical work of chemists into processes and products that can be used in manufacturing, although it is often difficult to determine where the work of one team member ends and the other begins. While chemists and chemical engineers both work with chemicals, chemists tend to work more with theoretical and small-scale reactions, while chemical engineers apply chemistry in large-scale manufacturing processes. The two areas differ in emphasis but are interdependent in manufacturing. This relationship is illustrated by the history of the rubber industry.

In the early part of the 20th century there was no need to define the word "rubber." There was only one rubber, natural rubber, and most books of that time simply defined rubber as "the milky product of the rubber tree." Today, the term "rubber" includes not only that natural product but also a host of synthetic rubbers and some plastics that exhibit certain "rubbery" characteristics. All types of rubber are composed of long, flexible polymer molecules that are able to be crosslinked to form a three-dimensional structure. Figure 2-5 shows a two-dimensional cross-section of a small part of such a molecule.

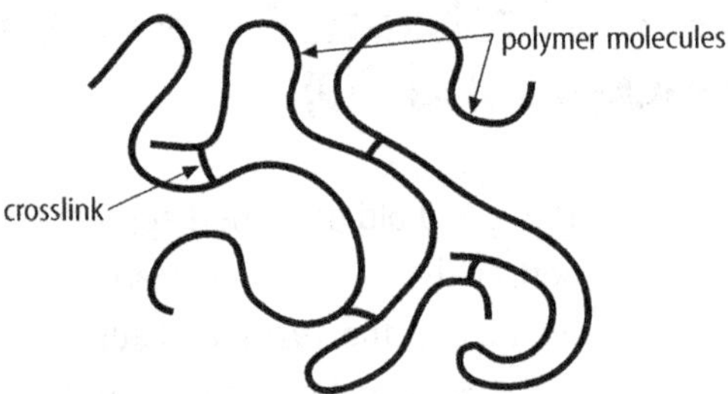

Figure 2-5: A two-dimensional cross-section of a crosslinked polymer

This crosslinked network is highly elastic, being able to stretch without breaking and return to its original form when the stretching force is removed. In order for rubber to have the characteristics desired in many industrial products, it must be crosslinked. Traditionally, this crosslinking involves heating raw natural rubber, which has very little crosslinking, with a crosslinking chemical such as sulfur in a process known as vulcanization. Chemists can vary the properties of the rubber over a wide range by varying the crosslinking chemical and the heating conditions. A second major way to modify properties of rubber is to introduce a reinforcing filler such as carbon black. Adding carbon black to rubber improves the abrasion resistance, resiliency, and strength of the compound. Our transportation systems depend on tires, and tires depend on carbon black to give strength and durability to the tire treads.

Once scientists discovered ways to improve the properties of rubber, chemical engineers could start developing a way to manufacture the product on a large scale. Incorporating an additive such as carbon black to a batch of raw rubber presented a unique mixing problem. The additive had to be uniformly distributed throughout the bulk of the rubber, and the particle size of the additive had to be small enough to give homogeneous properties. Carbon black in particular posed a difficult problem because its low density allowed it to float out of the raw rubber batch into the air, dirtying the surroundings as well as changing the proportions in the batch. The chemical engineer F.W. Banbury solved this problem by inventing a new type of mixer, thus changing the course of the rubber industry. Banbury's mixer was totally enclosed to contain dusty ingredients, and it produced an extremely uniform batch with a significant savings in time. In this heavy-duty mixer, the batch enters through the feed hopper door and is fed to the mixing cored rotors by a floating weight. Cooling sprays regulate the temperature during the mixing process, and the final mixed batch exits through the sliding discharge door. (See Figure 2-6.)

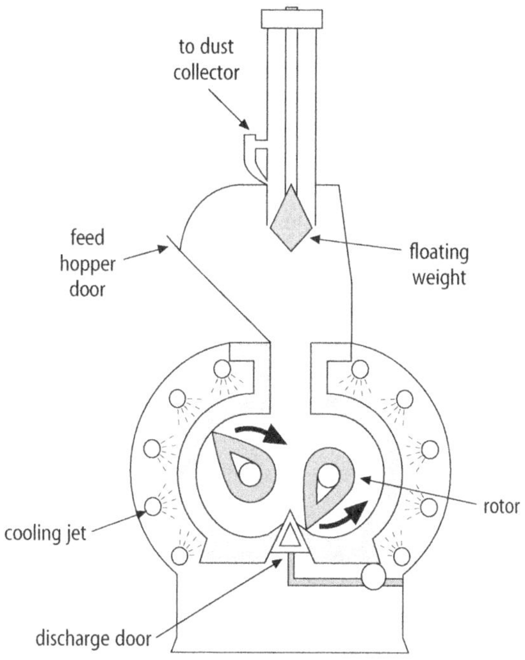

Figure 2-6: Schematic cross-section of a Banbury mixer

Since 1920, the Banbury mixer has been widely used in the rubber industry, and it became the prototype for all internal mixers. (See Figure 2-7.)

Figure 2-7: A Banbury mixer

Banbury's work linked scientists' discoveries about the properties of rubber with the need of industry to provide better tires for transportation. Even though the equipment needed for plant-scale production required millions of dollars of capital investment, the cost savings of even two to three cents per pound of product through better design or more efficient mixing made a tremendous difference in the profit margin of rubber companies. According to Jim Miller, an engineer at DuPont, the driving force of the chemical engineer's work is "to answer the seemingly simple question, 'What is the best way to do this job?' Whether a company makes money or loses it—and how much—is directly proportional to the quality of the engineering work."

● Mixing in Many Cases (Student Background 2E)

*"But you've no idea what a difference it makes, mixing it with other things—
such as gunpowder and sealing wax."*

—Lewis Carroll in *Through The Looking Glass*

Engineers regularly encounter a variety of mixing challenges. One useful way to
consider mixing problems is to study the phases (solid, liquid, or gas) that are
involved in a particular process.

Mixing of Liquids

Fluid mixing may be a simple, easily evaluated process or an extremely complex
one with many variables to consider. Fluid mixing includes both liquids and
gases. Generally, liquids are mixed by using some type of rotating blade called
an impeller. Motion in a liquid is largely dependent on three factors: the
container in which the liquid is being mixed, the impeller system, and the
nature of the liquid itself. Containers are usually cylindrical tanks. (In square
tanks it is too difficult to mix the liquid in the corners.) Two common types of
impeller are the pitched blade turbine, which is used in large tanks, and the
hydrofoil impeller, in which the blades are curved as well as twisted to give
greater flow and better mixing. (See Figure 2-8.)

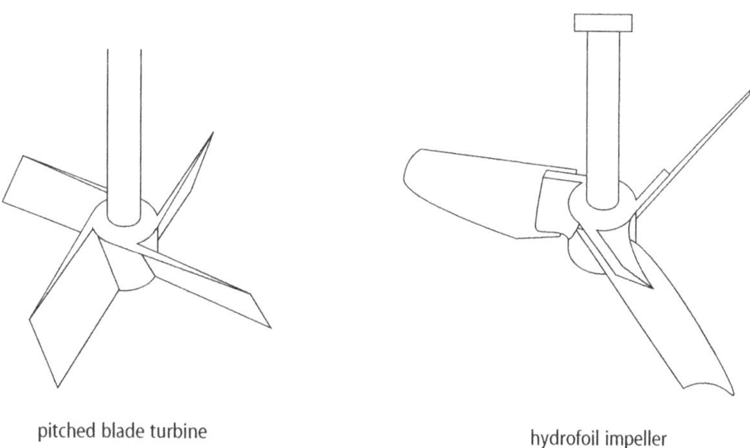

Figure 2-8: Common types of impellers

The movement of the impellers causes fluid motion within the liquid and, with
the walls of the tank, creates flow lines that mix the liquid. (See Figure 2-9.)
Baffles at the sides of the tank break up the swirling motion of the liquid and
create both vertical and horizontal flow patterns. Overall, impellers fall into two

basic categories: radial and axial. Radial impellers push liquid horizontally away from the impeller blades toward the tank wall. Axial flow impellers create flow vertically up or down away from the blades in a motion parallel to the shaft.

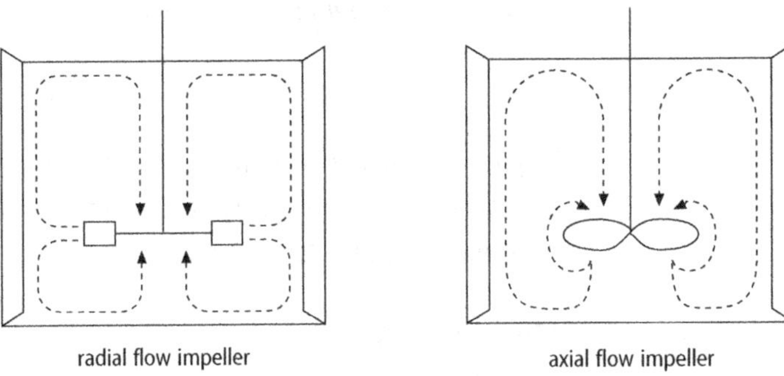

Figure 2-9: Mixing flow patterns in liquids

Solid-Liquid Mixing

Solids may need to be mixed with a liquid for many reasons, such as to dissolve the solid, to promote a chemical reaction between a solid and a liquid, or to produce a homogeneous suspension medium for processing. In this case the solid particles are kept in motion by currents in the liquid. The efficiency of the mixing depends on the size, shape, and density of the solid particles; these qualities affect not only the particles' tendency to settle out of the liquid but also the viscosity and density of the liquid. Small, low-density particles easily follow the circulation pattern of the liquid, but larger, heavier particles tend to settle out against the fluid motion.

Gas-Liquid Mixing

Mixing between gases and liquids may need to take place in a variety of industrial operations where the purpose of the mixing is to disperse gas bubbles in the liquid phase. In some cases, gases and liquids are mixed to provide a stable batter or foam, such as in many food industry processes, or to generate a froth for flotation of solid particles, such as in many metallurgical processes. Several techniques are used to introduce and mix gases with liquids. Gas bubbles may be generated directly within the solution by the decomposition of a carbonate or bicarbonate to liberate carbon dioxide bubbles. Mechanical devices such as the ones shown in Figure 2-10 provide alternative methods of introducing gas bubbles.

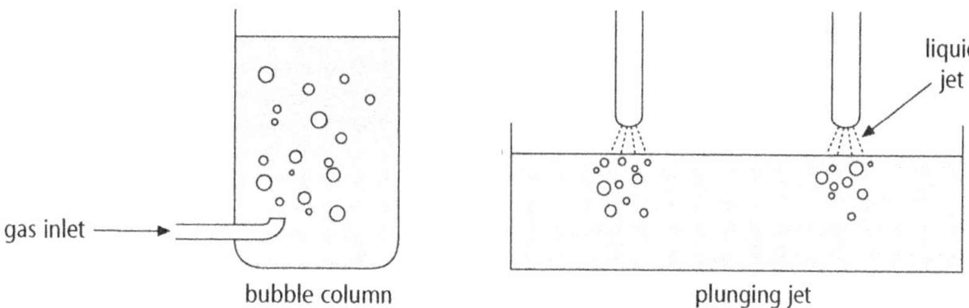

Figure 2-10: Mechanical devices can be used to introduce gas bubbles.

Mixing of Solids

Many industrial operations involve only solid-solid mixing with no liquid or gas phase present. Here a unique problem known as segregation presents itself. Segregation is the tendency of solid particles to separate out according to their size and/or density. Student Activity 1 illustrates a real-world case of segregation in a box of raisin-bran cereal. In practice, there is almost always some difference in the physical properties of the solids used in a mixture, so choosing an effective method for mixing solids requires an understanding of the phenomenon of segregation. Of all phase combinations, the mixing of different solids is the hardest to define and to evaluate. Generally, a specific system must be set up for each case. The most common types of solid mixers are either tumbler mixers or ribbon mixers (see Figure 2-11), with many patented designs on the market for both types.

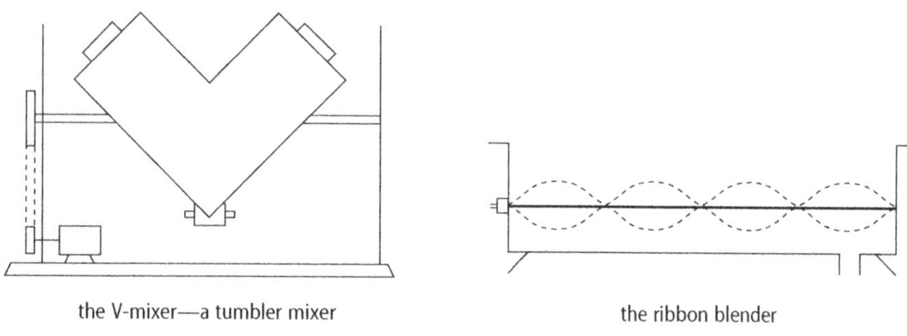

Figure 2-11: The most common types of solid mixers

The tumbler mixer consists of a totally enclosed vessel that rotates about an axis, causing the particles of the solid batch inside to "tumble" over each other on the surface of the mixture. Ribbon mixers operate within a stationary trough or open cylinder, and the solid particles are relocated by the moving ribbon.

● Overhead 2-2: Common Types of Impellers

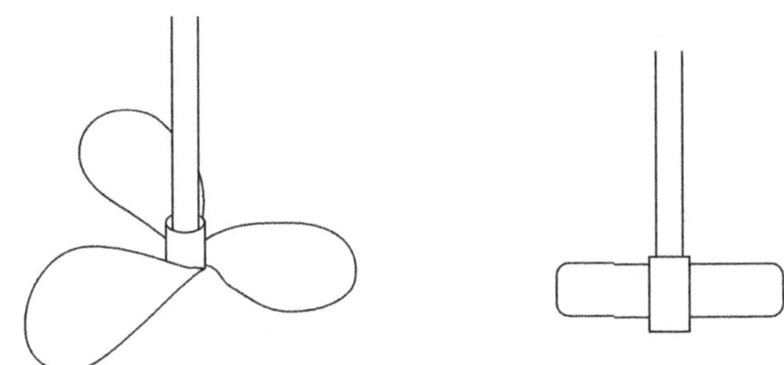

three-bladed marine type

simple paddle

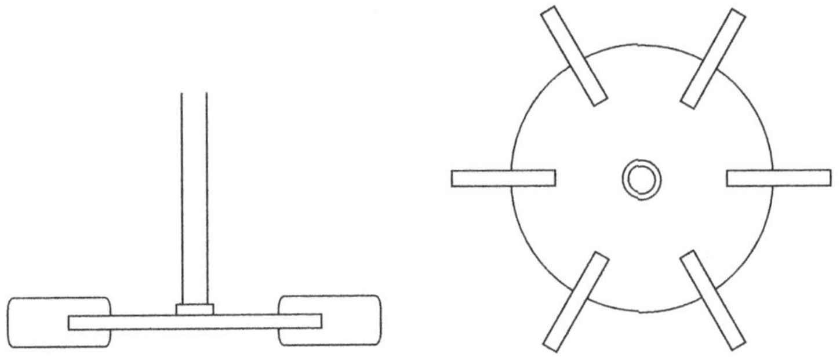

six-bladed disc turbine (side view and top view)

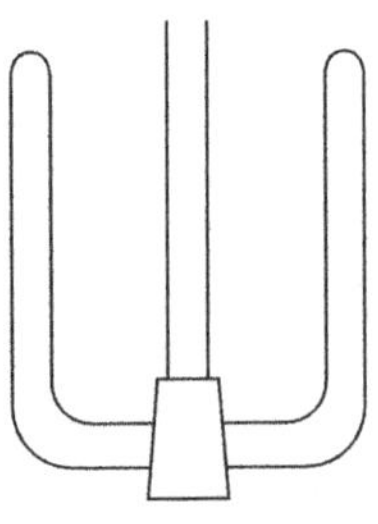

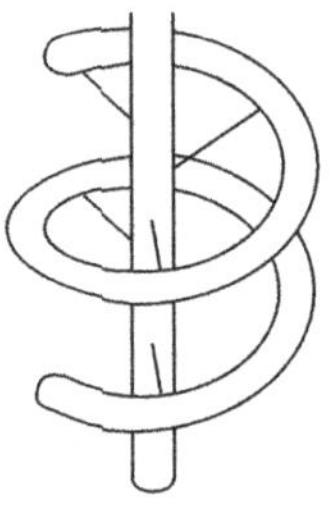

anchor impeller

helical ribbon

● Overhead 2-3: Typical Arrangement for a Mechanical Mixer

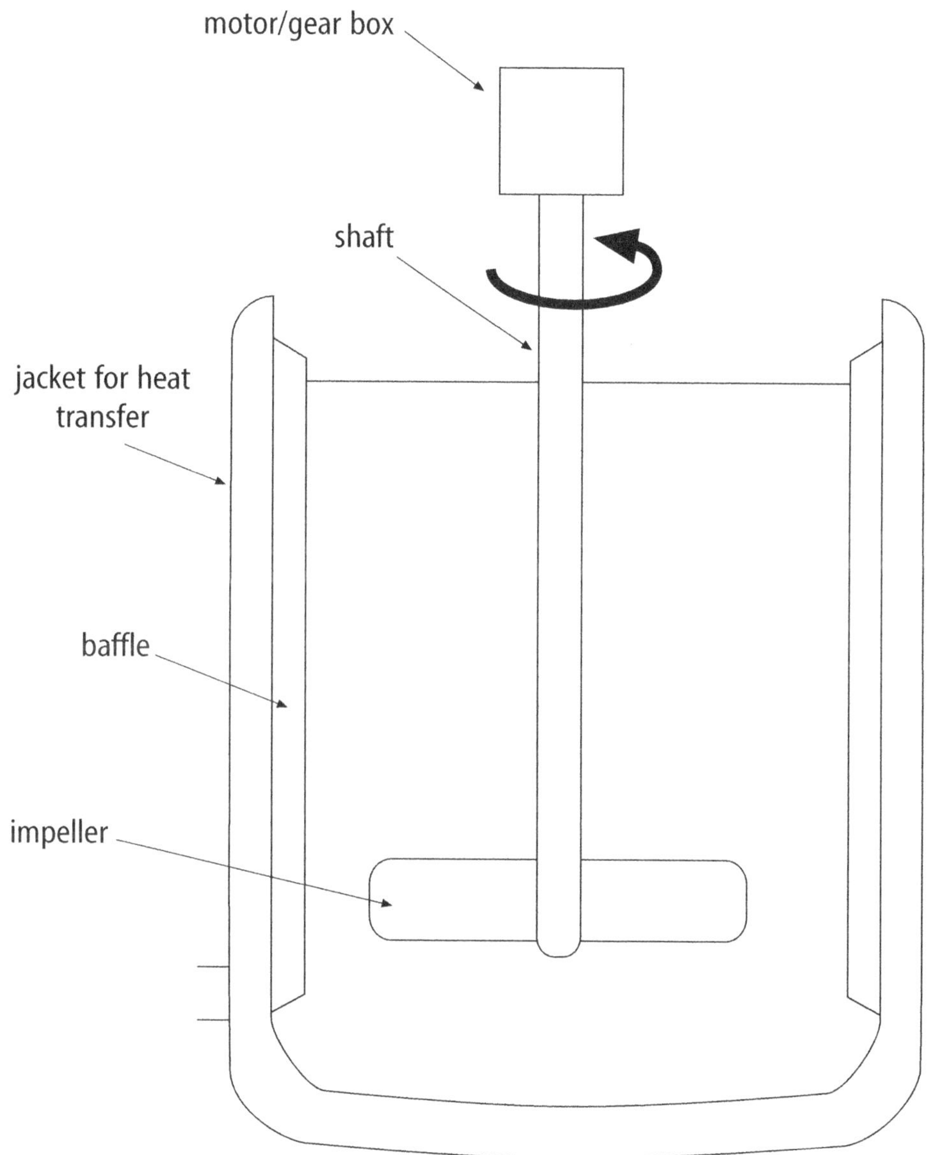

Section 3

Why Do Particles Segregate?

● Overhead 3-1: Some Possible Particle Distributions in a Two-Component Solid Mixture

total segregation—no mixing

random mixing

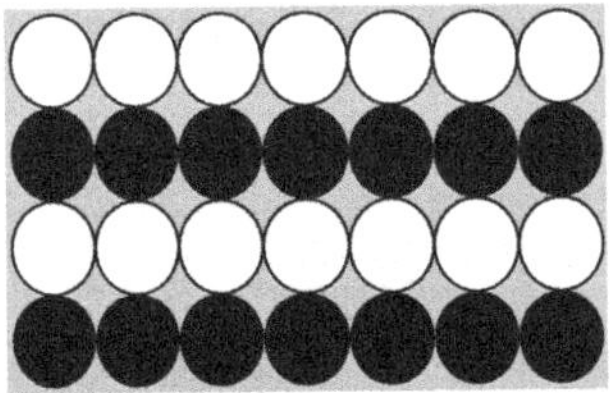

fully mixed—not random

● Why Do Patterns Appear? (Student Activity 3A)

For solid mixtures, the ability to form a homogeneous mixture depends on the nature of the batch ingredients. Segregation is the natural tendency of a solid mixture to separate into various components due to differences in size, density, or shape of the particles. Smaller, heavier, smoother, rounder particles sink through larger, less dense, or more jagged ones. The ability of particles to flow becomes an important factor whenever the mixture is poured. An example from industry is when the mixture of raw ingredients that will be melted together to form glass flows from the holding bins into the glass furnace.

Novelty and toy stores often market applications of segregation in solid mixtures under names such as Sandsations or Magic Windows. In the following activity you will construct a similar segregation system and investigate its properties.

Safety and Disposal

No special safety or disposal procedures are required.

Materials

Per group of 2 students
- 2, 9-cm plastic Petri dishes with lids
- modeling clay
- 10.0 g colored sand
- 2.5 g finely ground table salt
- 3.5 g sugar
- 5.5 g colored crystal sugar

Procedure

❶ Invert the lid of one of the Petri dishes so that the dish can be set inside it with the underside of the Petri dish against the inside of the lid.

❷ Remove the Petri dish and place a pencil-thin roll of modeling clay along the inner edge of the lid. (See Figure 3-1.)

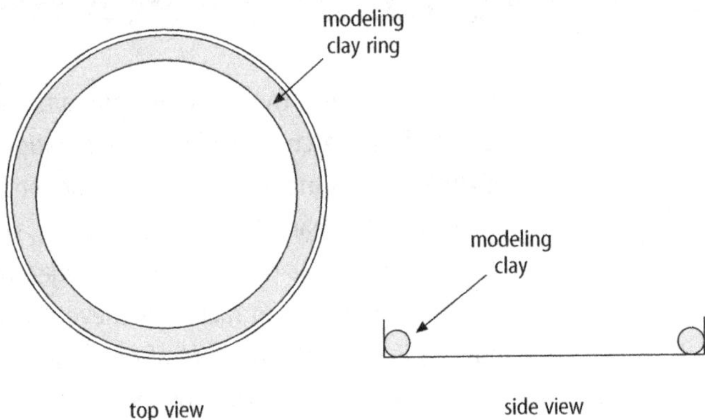

Figure 3-1: Place a modeling-clay ring inside the lid of the Petri dish.

3 With the lid flat on the table, place the sand and salt in the center of the lid so that it is not in contact with the modeling clay. Set the Petri dish onto the modeling clay ring so that a space of approximately 0.5 cm remains between the underside of the Petri dish and the inside of the lid, forming a sandwich with the sand and salt in the center.

4 Press firmly so that the modeling clay will hold the Petri dish firmly to the lid.

5 Repeat steps 1–4 using the second Petri dish and lid set but filling this one with the two sugar samples.

6 Rotate, turn, and shake the pattern windows gently and observe the movement of the various components in the mixtures.

Summary Questions

1 Describe in detail the patterns formed by the mixtures as you rotated, turned, and shook the windows.

____

____

____

____

____

❷ Is it possible to make the mixtures in the pattern windows homogenous? Explain.

____

____

❸ In these systems, what appears to be the major factor influencing the segregation of the particles in the mixtures?

____

____

____

____

Materials Notes

Crystallized sugar is available in the baking supplies section of most supermarkets with other cake-decorating supplies. The colored sand used in this activity is available at hobby or craft stores. The amounts of sand, salt, and sugars are not critical, but the relative proportions listed give good results.

Getting Ready

Any brand of salt may be used; you can grind it ahead of time in a mortar and pestle until it is quite fine. Always wear goggles when grinding samples.

Procedure Notes

Students should roll a pencil-thin "snake" of modeling clay and place it along the inside edge of the Petri dish lid. When the Petri dish is placed right-side-up in the lid, the modeling clay will provide a good seal and students may rotate the window with no loss of contents. A band of clear tape may be wrapped around the outside to ensure that the two halves of the Petri dish do not come apart.

Answers to Summary Questions

❶ *Describe in detail the patterns formed by the mixtures as you rotated, turned, and shook the windows.*

Answers will vary.

❷ *Is it possible to make the mixtures in the pattern windows homogeneous? Explain.*

It is impossible to achieve homogeneity with these systems.

❸ *In these systems, what appears to be the major factor influencing the segregation of the particles in the mixtures?*

The difference in particle size appears to cause the segregation.

● Will Grinding a Sample Improve Homogeneity? (Student Activity 3B)

Grinding a solid mixture to a uniform particle size is a standard industrial procedure. In this activity you will study the effect of grinding on homogeneity.

Safety and Disposal

Always wear goggles when grinding samples. No special disposal procedures are required.

Materials

Per group of 2 students
- 3.5 g finely ground sugar
- 5.5 g colored crystal sugar
- mortar and pestle
- 9-cm plastic Petri dish with lid
- modeling clay
- goggles

Procedure

❶ Mix the two sugar samples together and grind them vigorously in the mortar and pestle for 2 minutes.

❷ Invert the lid of one of the Petri dishes so that the dish can be set inside it with the underside of the Petri dish against the inside of the lid.

❸ Remove the Petri dish and place a pencil-thin roll of modeling clay along the inner edge of the lid. (See Figure 3-2.)

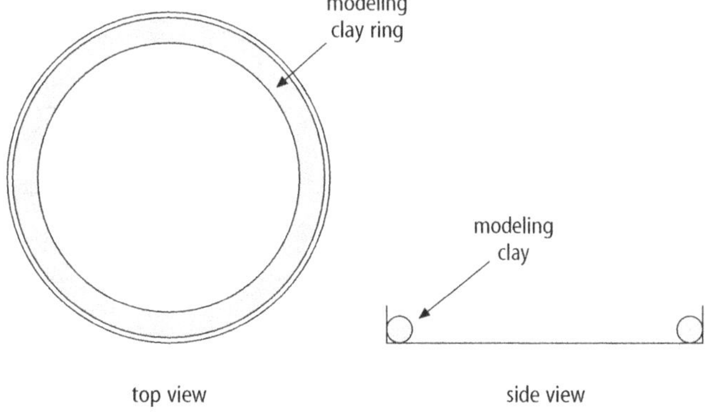

Figure 3-2: Place a modeling-clay ring inside the lid of the Petri dish.

❹ With the lid flat on the table, place the sugar in the center of the lid so that it is not in contact with the modeling clay. Set the Petri dish onto the modeling clay ring so that a space of approximately 0.5 cm remains between the underside of the Petri dish and the inside of the lid, forming a sandwich with the sugar in the center.

❺ Press firmly so that the modeling clay will hold the Petri dish firmly to the lid.

❻ Rotate, turn, and gently shake the window to observe the movement of the various components.

Summary Questions

❶ How much does the ground sample segregate in the pattern window? Compare the results with the two-sugar pattern window constructed in Student Activity 3A. Did grinding the sample appear to improve its homogeneity?

____

____

____

____

____

____

❷ Can you suggest a reason for this?

____

____

____

____

____

Instructor Notes for Student Activity 3B

Materials Notes

Crystallized sugar is available in the baking supplies section of most supermarkets with other cake-decorating supplies. The amounts of the two sugars are not critical, but the relative proportions listed give good results.

Procedure Notes

Students should roll a pencil-thin "snake" of modeling clay and place it along the inside edge of the Petri dish lid. When the Petri dish is placed right-side-up in the lid, the modeling clay will provide a good seal and students may rotate the window with no loss of contents. A band of clear tape may be wrapped around the outside to ensure that the two halves of the Petri dish do not come apart.

Answers to Summary Questions

❶ *How much does the ground sample segregate in the pattern window? Compare the results with the two-sugar pattern window constructed in Student Activity 3A. Did grinding the sample appear to improve its homogeneity?*

Much less segregation occurs in the ground sugar samples than in the unground sample in Student Activity 3A. It appears that grinding the mixture significantly improved its homogeneity.

❷ *Can you suggest a reason for this?*

When the sample is ground, the particles of the two kinds of sugar are reduced to a similar size. The more similar the size of the particles, the less segregation will occur and the more homogeneous the mixture will remain.

● Will Other Solids Segregate? (Student Activity 3C)

From the baker stirring muffin ingredients in the kitchen at home to the chemist mixing reagents with a stirring rod in a beaker, to the chemical engineer trying to ensure efficient contact of raw materials in a large industrial plant operation, mixing is an important part of our lives. Solid-solid mixtures, as we have seen, are subject to the special problem of segregation. In this activity you will investigate this problem further by setting up several two-component systems and attempting to determine the major causes of segregation in those systems.

Materials

Per group of 2 students

- variety of containers with lids
- variety of solid components
- scoopula
- small funnel

Procedure

Using the containers and materials provided, experiment with as many two-component systems as possible to investigate the problems involved in achieving homogeneity in a system. Write a procedure for your experiment and construct a data table to summarize your findings.

Summary Questions

❶ In which combinations was it most difficult to achieve a homogeneous mixture? In which cases was it fairly easy? Why do you think this was the case?

____

____

____

____

____

____

② Did the shape and type of container significantly affect the mixing process? Explain.

③ Did the amount and type of agitation significantly affect the segregation? Explain.

④ Using evidence from your investigations in this activity, discuss what factor you found to have the most influence in causing a mixture to segregate.

Students will experiment with as many two-component systems as time and materials permit in order to determine the factors that cause a mixture to segregate and to observe various patterns of mixing.

Materials Notes

Assemble a sizable collection of jars, bottles, vials, and Petri dishes of various shapes and sizes and a large number of different common materials, such as dried beans, rice, macaroni, split peas, marbles, colored sand, silver dragées, and decorative sugar pellets (nonpareils) that can be used for mixtures.

For advanced students, use small magnets to introduce the concept of polarity.

Sample Data Table

Sample Data for "Will Other Solids Segregate?"		
Components	Process	Results
red beans/rice in a glass jar	rotated jar for 30 seconds	mixture slid as a mass, did not mix
	shook jar gently for 30 seconds	beans rose to top of mixture

Answers to Summary Questions

❶ *In which combinations was it most difficult to achieve a homogeneous mixture? In which cases was it fairly easy? Why do you think this was the case?*

Combinations in which the size of the particles was very different, such as marbles and rice or nonpareils and beans, segregated markedly, with the smaller particles always sifting to the bottom of the container. If the components in the mixture had roughly the same particle size, such as with two kinds of dried beans, good mixing could be achieved by rotating or shaking the container.

❷ *Did the shape and type of container significantly affect the mixing process? Explain.*

Answers will vary with the shape and type of container, but the major factor will always be the particle size.

❸ *Did the amount and type of agitation significantly affect the segregation? Explain.*

Answers will vary.

❹ *Using evidence from your investigations in this activity, discuss what factor you found to have the most influence in causing a mixture to segregate.*

It appears that relative particle size of the solids in a mixture is a major factor in causing segregation. (See examples cited above.)

● A Particle Race (Challenge Activity)

Use your understanding of the movement of particles with respect to each other to win the race!

Materials

Per team of 4 students

- 30–40 cm clear rigid plastic tubing (1.5- to 2.5-cm inner diameter) with tape bands
- 2 solid rubber stoppers
- small funnel
- sand (enough to fill the tube two-thirds full)
- metal ball (slightly smaller than the diameter of the tube so that it can move freely)

Procedure

❶ Stopper the taped end of your plastic tube tightly.

❷ Using the small funnel, pour sand up to the top of the middle band of tape.

❸ Drop the metal ball into the tube on top of the sand.

❹ Pour sand through the funnel onto the top of the ball. Continue until the sand reaches the top of the upper band of tape.

❺ Stopper the other end of the tube tightly.

❻ Within 2 minutes, devise a strategy to move the metal ball from the middle of the sand column to the top of the sand as quickly as possible without opening the tube.

 *You may not handle the tube while planning your strategy.*

❼ At the end of your 2-minute strategy session, appoint one member to carry out the task. At a signal from the instructor, the designated team member should pick up the tube and attempt to move the ball according to the team's strategy. Other team members may suggest new strategies as the race progresses if their original plan is not successful. The first team to move its metal ball to the top of the sand column wins the race.

Discussion

What was the method used by the winning team? Why did it work? Other teams should experiment with this method and practice modifications. When each team has had a chance to practice the technique, the race may be repeated.

Chemical Manufacturing: The Process of Mixing

This activity may be used to assess understanding of the concept of segregation. If students have understood that, in general, relative size is the determining factor in segregation and have concluded from their experimental activities that larger particles will generally segregate to the top of a mixture with the smaller particles sifting to the bottom, they should conclude that shaking the tube vertically will allow the larger ball to rise slightly each time, allowing the smaller sand grains to flow beneath it, gradually raising the ball to the top of the sand.

Materials Notes

Large marbles may be substituted for the metal balls. Use clear, rigid plastic tubing for this activity. It is inadvisable to substitute glass tubing due to the danger of breakage during an enthusiastic race. Prepare tubes by placing three tape bands around the tubes at regular intervals as shown in Figure 3-3.

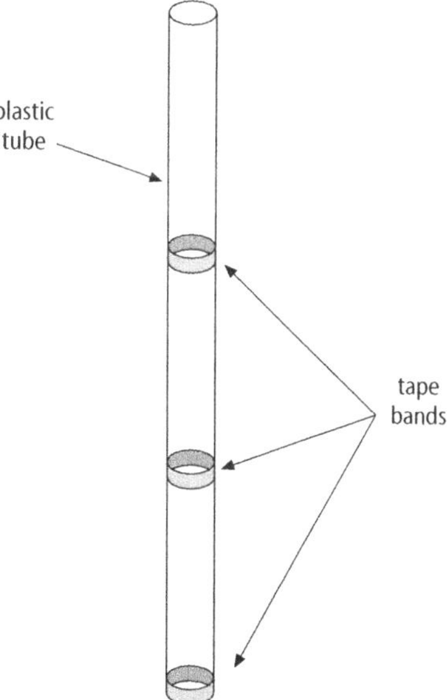

Figure 3-3: Place three tape bands around each tube at regular intervals.

A commercial tube and ball set already prepared may be purchased from Educational Innovations, Inc. (888/912-7474), under the name of Magic Sand Wand.

Section 4

Where Does the Bounce Come From?

● This Is a Chemical Industry: The Dow Story (Student Background 4A)

How did a small-town company in the Midwestern part of the United States become one of the world's largest chemical complexes? The unlikely cornerstone of the story is a 24-year-old man, a new graduate of Case School of Applied Science (now Case-Western Reserve University) whom the locals dubbed "Crazy Dow" when he first appeared in Midland, MI. Only Herbert Dow and a few investors believed that his secret process for extracting bromine, and later chlorine, from the salt brine that lay beneath the ground of the Saginaw Valley could compete with the powerful German syndicate Deutsche Bromkonvention, which dominated the world market for the bromides used in photography and medicines. In 1889 Dow sketched in one of his notebooks a diagram accompanied by this inscription: "to make a cell (electrolytic) for the decomposition of bromides. Have the plates horizontal and sealed into the sides of wooden boxes for cells." While his idea for the use of electrolysis in a chemical production was not new, his uniquely simple cell design eventually allowed the Dow Chemical Company, founded in 1897, to produce bromine and chlorine in large quantities, laying the groundwork for an American chemical industry.

By 1914 Dow Chemical Company was producing a variety of chemicals. With the beginning of World War I in Europe, Dow foresaw a serious world shortage of dyes used by textile manufacturers because Germany had been supplying nearly 80% of the dyes sold on the world market. Experiments were begun to synthesize indigo, the most widely used dye, and within 18 months the first synthetic indigo ever made in the United States was shipped from Dow's Midland plant to a North Carolina textile plant.

During World War II, the federal government urgently needed massive quantities of magnesium metal for construction of lighter, faster war planes. Dow Chemical Company, which had perfected a process of extracting magnesium metal from brine and had continued to research and produce the material even though the market was unprofitable, was able to supply the needed metal and provide the expertise to other companies to begin production. Also, when the Japanese armies captured the rubber plantations in the Far East, causing an acute shortage of tires, Dow chemists developed styrene and butadiene, essential ingredients for making synthetic rubber, and shared the raw materials and the process with other companies so that the chemical industry could supply a synthetic substitute for the unavailable natural rubber.

Dow Chemical Company research has expanded into many fields, including polymers of many kinds, adhesives, ceramics, and the production of many chemicals. Dow's Saran® and Ziploc® products have become household words, as have the "scrubbing bubbles" that advertise a line of household cleaners. Dow Plastics is a major manufacturer and supplier of polyethylene and polystyrene products and of engineering plastics. Products from Dow Plastics are used in a variety of applications, including automotive, durable goods (electronics, appliance, building and construction, and medical), and packaging. From its beginning in a single wooden shack in Midland, MI, the company has expanded its operations from coast to coast and into the international market, producing and marketing chemicals in the United States, Canada, Latin America, Europe, and the Pacific.

Where Does the Bounce Come From?
(Student Background 4B)

"Bounce," or the property of elasticity, characterizes certain types of polymers. Polymers ("poly" meaning "many" and "mer" meaning "units") are giant molecules, or macromolecules, that have been made by chemically bonding together great numbers of smaller units called monomers ("mono" meaning "one"). A single polymer molecule typically contains anywhere from 1,000 to 50,000 monomer segments bonded together. Polymers can be made of one type of monomer exclusively, or they can contain several types of monomers. The conditions needed to chemically bond these monomers together may include use of high pressure, high temperature, and/or the presence of a catalyst.

Depending on the conditions used in the polymerization reaction, these monomers can be joined to form linear polymers and branched polymers. Figure 4-1 shows simplified diagrams of these types of polymers.

Figure 4-1: Diagrams of linear and branched polymers

Different types of polymers have different properties. Some polymers are very hard and rigid, while others are soft and flexible. Some polymers are resistant to heat, while others melt easily. Some polymers can be molded and shaped into useful objects, while others resist deformation, even at high pressures and temperatures. The properties of polymers depend on many factors such as the type of monomer, the molecular weight distribution, the amount of branching, the strength of the intermolecular forces, the regularity of the monomer units, and the flexibility of the polymer chain itself.

Because of the extreme length of most polymer chains, they frequently exist as tangled coils, like a wad of tangled wire. At times this three-dimensional coiled structure is stabilized by attraction between different regions of the chains (as with the alpha-helix structure that exists in a molecule of DNA, a naturally occurring polymer). This three-dimensional structure can also be stabilized by the addition of a crosslinker. A crosslinker is a small molecule or ion that holds two (or more) different polymer molecules together and thus restricts the movement of the individual polymer molecules in the sample.

Stress to the coiled polymer molecule may cause the three-dimensional coil to unwind or stretch out. However, once the stress is removed, the molecule has a tendency to return to the random coiled shape; hence the elastic nature of some polymers.

White glue (such as Elmer's® Glue), which is used in Part I of Student Activity 4, contains the polymer polyvinyl acetate; the repeating monomer of this polymer is shown in Figure 4-2. The subscript "n" stands for an unspecified (but very large) number, indicating that a large number of these monomers can be linked to form the long polymer chain. The wavy lines also designate repeating monomer units on the chain.

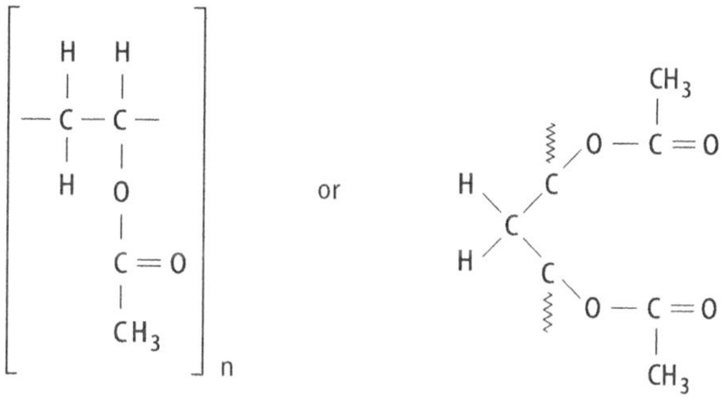

Figure 4-2: Repeating monomer of polyvinyl acetate

Crosslinking these polyvinyl acetate molecules with the borate ion $B(OH)_4^-$, as shown in Figure 4-3, causes the glue to assume a much more rubbery and bouncy consistency because the new structure has more limited freedom of motion. Borax is a common source of the borate ion.

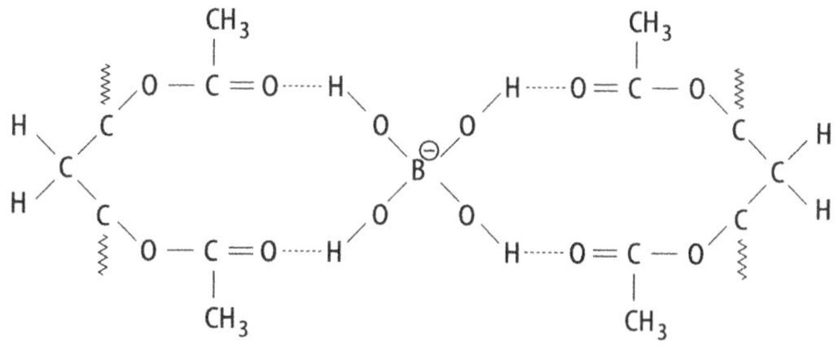

Figure 4-3: Possible hydrogen bonding sites on polyvinyl acetate

Polyvinyl alcohol, which is used in Part II of Student Activity 4, contains the repeating monomer shown in Figure 4-4. Again, note that the large value of "n" denotes polymer chains that may be several thousand atoms long.

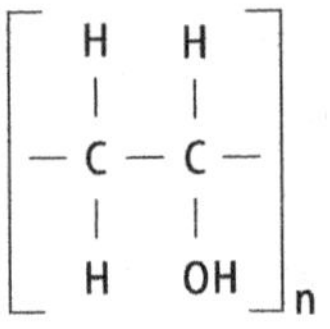

Figure 4-4: Repeating monomer unit of polyvinyl alcohol

The borate ions $B(OH)_4^-$ present in the borax solution can cross-link with the hydroxy groups of the polyvinyl alcohol in one of two ways. (See Figures 4-5 and 4-6.)

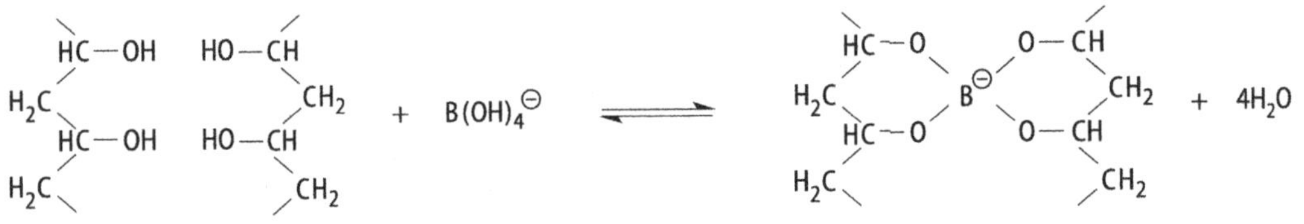

Figure 4-5: Crosslinking of polyvinyl alcohol with the borate ion

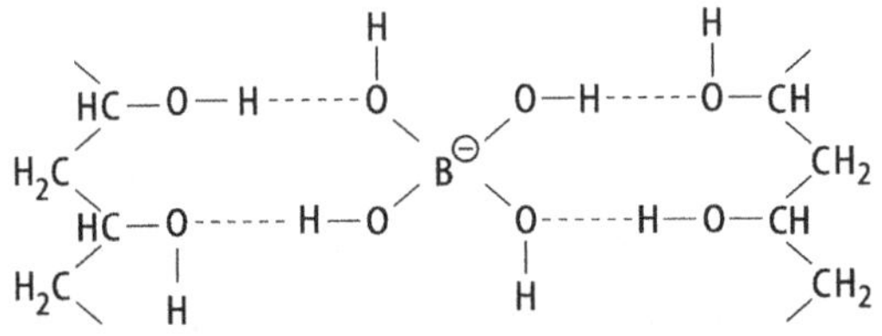

Figure 4-6: Possible hydrogen bonding between polyvinyl alcohol and the borate ion

While the hydrogen bonding shown in Figure 4-6 is a weaker attraction than the covalent crosslinks shown in Figure 4-5, it does play a role in the gel's formation. The crosslinking and bonding shown in Figures 4-5 and 4-6 result in a three-dimensional network of polymer chains connected by borate. Water molecules occupy most of the space within the three-dimensional network that comprises the gel.

● Table of Common Polymers (Student Background 4C)

Polymer	Monomer Unit	Properties	Polymer Uses
polyethylene	$-CH_2-CH_2-$	lightweight, tough, flexible; resistant to water, frost, electricity, and chemicals; easily colored and molded; inexpensive to produce	low-density: bowls, buckets, beakers, toys, squeeze bottles and tubes, stoppers for bottles, piping, power and telephone cable, packaging, and disposal bags linear-low-density: trash and merchandise bags, large hollow containers, housewares, traffic control devices, wire and cable applications
polypropylene	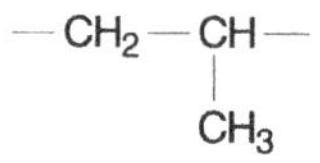 $-CH_2-CH-$ with CH_3 branch	similar to polyethylene except more rigid and more heat-resistant	hospital and lab equipment that needs to be sterilized, baler twine and rope, chair seats, tool handles, shoe heels, toys
polyvinyl chloride	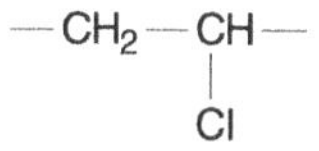 $-CH_2-CH-$ with Cl branch	strong, water- and weatherproof, durable	pipes for chemical plants, water, gutters, and soil; boots and casual shoes; washable vinyl wallpaper; tablecloths; shower curtains; baby pants; inflatable toys and squeaky dolls; floor tiles
polystyrene	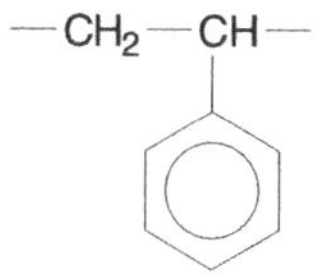 $-CH_2-CH-$ with phenyl (benzene ring) branch	crystal clear but brittle; if stretched becomes opaque; good insulator; resistant to water and acid	clear storage containers, reels and spools for film or tape, yogurt containers, disposable cups, refrigerator linings and trays, TV and radio cabinets, pen barrels expanded polystyrene: ceiling tiles, insulation, meat and food trays, packaging
acrylic (polymethyl methacrylate)	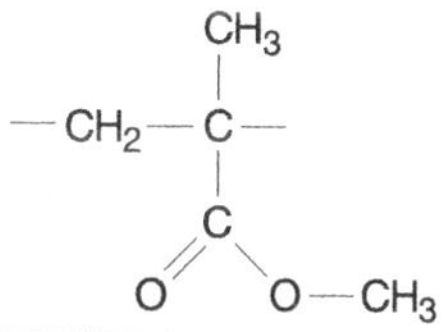 $-CH_2-C-$ with CH_3 branch and $C(=O)O-CH_3$ ester group	hard, transparent, can be colored, weatherproof, lightweight, easily molded	replaces glass where unbreakable or curved window is required, watch glasses, contact lenses, false teeth, goggles, safety guards on machines, street light covers, car rear lights
nylon	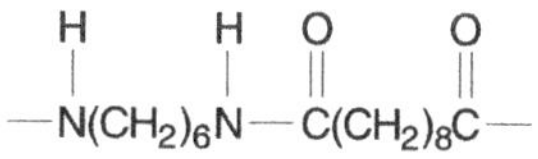 $-N(CH_2)_6N-C(CH_2)_8C-$ with H, H, O, O groups	very tough and rigid; resistant to water, corrosion, wear, and high temperatures	solid: used in gears, bearing parts, curtain runners thick thread: fishing nets, tennis racket strings, ropes, and carpets thinner thread: bristle for brushes, fishing lines fine thread: used widely in textile industry
polyester	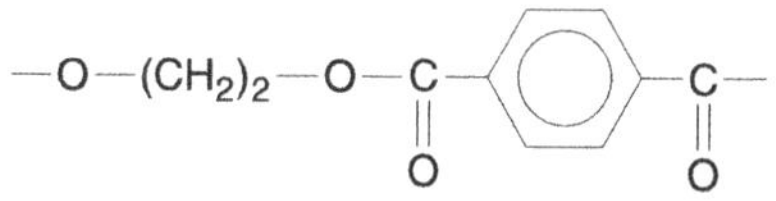 $-O-(CH_2)_2-O-C-$ (benzene ring) $-C-$ with O, O groups	does not stretch; keeps shape well; brittle but, in combination with glass fibers, very strong	transparent roasting bags, magnetic tape, glass fiber reinforcement for boat hulls, some car bodies, some roofing panels, resistant finish for furniture against chemicals and heat
polyurethane	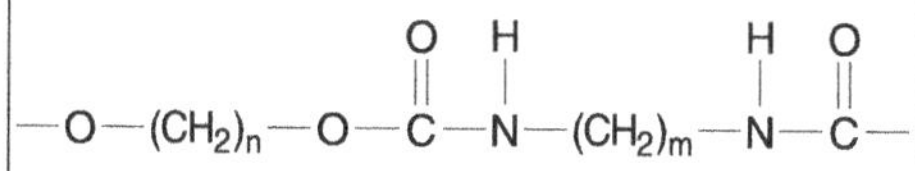 $-O-(CH_2)_n-O-C-N-(CH_2)_m-N-C-$ with O, H, H, O groups	rigid foam: heat insulator, water-resistant flexible foam: sound absorber, insulator	surface coating for furniture and floors, foams used to form to furniture before covering or shipping, machine parts, stools, bowls, weatherproof sealing strips

● How Do We Know If It's Mixed Enough? (Student Activity 4)

According to the old saying, "The proof of the pudding is in the eating," and without a doubt this is true of the mixing process in a manufacturing operation. The ultimate way to assess the quality of the mixing process is to look at the quality of the resulting product. If a rubber tire blows out because of weak spots or if a glass windowpane is marred by a wavy appearance or small bubbles, the problem may lie in improper mixing of ingredients during the manufacturing process. Mixing quality may be judged by the homogeneity of a sample, such as in the pattern windows experiment. When a chemical reaction is involved, it may be judged by such factors as reaction yield and the properties of the resulting product.

In this experiment you will prepare samples of several crosslinked polymers, using different mixing times for each sample, and compare the properties of the product obtained in each case.

Safety

Goggles are mandatory; wear them at all times during this experiment.

Materials

Per group of 2 students

- 3 small paper cups
- stirring rod
- 16.5 g white school glue
- 16.5 g polyvinyl alcohol solution
- 30 mL 4% borax solution
- stopwatch or clock with a second hand
- graduated cylinder or small metric measuring cup
- acetate sheet or wax paper

Per class

- balance (triple beam or electronic)

Procedure for Part I: Borax/Glue (polyvinyl acetate) Crosslinked Polymer

1. Weigh out 5.5 g glue directly into one of the paper cups on the balance.

2. Add 5 mL 4% borax solution to the glue while stirring and continue to stir the mixture for a total of 30 seconds.

Chemical Manufacturing: The Process of Mixing

❸ Spoon the mixture from the cup onto an acetate sheet and label it with the time of mixing.

❹ Repeat steps 1–3 using a clean paper cup and a clean spoon, this time stirring the mixture for a total of 2 minutes before placing it on the acetate sheet.

❺ Repeat steps 1–3 a third time, but with this sample, after the stirring has produced a polymerized mass that clings to the stirring rod, remove the mass and work it with your fingers for a total mixing time of 8 minutes.

Procedure for Part II: Borax/Polyvinyl Alcohol Crosslinked Polymer

Repeat Part I, substituting polyvinyl alcohol solution for the glue in each of the trials.

Summary Questions

❶ Compare the properties of the three samples in Part I and the three samples in Part II as a function of the mixing time.

____

____

____

____

____

____

____

____

____

❷ From your observations of the products, would you conclude that the chemical reaction had been completed in all cases? Why or why not?

____

____

____

____

____

____

❸ From your understanding of crosslinked polymers, discuss changes in properties, such as changes in elasticity, that you observed as you kneaded the samples for 8 minutes.

____

____

____

____

____

____

____

____

____

Instructor Notes for Student Activity 4

Safety

Goggles must be worn at all times during this experiment.

Some people are allergic to borax. As a result, care should be taken when handling. Avoid inhalation or ingestion. Use proper ventilation when preparing the borax solution and wash hands after contact with the solid. Generally there is little danger when working with the crosslinked polymers, but people with sensitive skin or allergic reactions to detergents should avoid handling the polymers. They may carry out the 8-minute mixing by placing the polymer in a small plastic bag and kneading it in the bag.

If students wish to take the 8-minute products with them, the products may be stored in a plastic bag so that they will not dry out. If the product should spill on the carpet, apply vinegar to the spot and follow with a soap-and-water rinse. Do not allow the polymer to harden on the carpet. Caution students not to set the polymers on wooden furniture because the polymers will leave a water mark.

Getting Ready

Prepare the 4% borax solution by mixing 40 g laundry borax (sodium tetraborate decahydrate, $Na_2B_4O_7 \cdot 10\ H_2O$) in 1 L distilled water while stirring. Students may be interested in preparing their own borax solution using commercial 20 Mule Team Borax. Alternatively, liquid laundry starch that contains borax may be substituted for the borax solution.

The 4% polyvinyl alcohol solution is available commercially from chemical supply companies (Flinn Scientific, Batavia, IL) or may be prepared from the solid by using one of the following methods.

- Dissolve 40 g polyvinyl alcohol slowly into 1 L water while heating the mixture on a hot plate over moderately high heat and stirring constantly. The solution will initially be quite milky in color but will become clear when the polyvinyl alcohol is completely dissolved. The process may take about 30–45 minutes. Cool the solution before using it. If a slimy or gooey layer appears on the top during cooling, simply skim it off and discard.

- Dissolve 40 g polyvinyl alcohol in 1 L water in a microwave-safe container. Stir the solution and place it in a full-size microwave. Heat the solution for 8 minutes, stirring every 1–2 minutes. Do not attempt to make more than 1 L at a time.

- Cut one PVA (polyvinyl alcohol) film bag into small pieces and dissolve the pieces in 875 mL hot water (approximately 60°C or 140°F). Let the solution cool to room temperature before use. PVA laundry bags can be obtained from your local hospital. They are sold commercially in cases of 100 by the Associated Bag Company, 400 West Boden Street, Milwaukee, WI 53207-7120, 800/926-6100 (26-inch x 33-inch PVA bags, 1 Mil thick, Stock Number 19-1-05). A less expensive source is American Science and Surplus, 3605 Howard Street, Skokie, IL 60076, 708/982-0870, but these bags are not always available.

Answers to Summary Questions

❶ *Compare the properties of the three samples in Part I and the three samples in Part II as a function of the mixing time.*

The samples that are mixed for only 30 seconds consist of a clump of fairly solid material in the center and a good bit of watery material around it. The 2-minute samples are more completely solidified but appear to be quite soft and sticky. The glue/borax sample that was kneaded for 8 minutes is very firm and no longer sticky. The 8-minute polyvinyl alcohol/borax sample, while more pliable and less elastic than the glue/borax sample, is now a cohesive mass that is no longer sticky.

❷ *From your observations of the products, would you conclude that the chemical reaction had been completed in all cases? Why or why not?*

It appears that the crosslinking is incomplete if the sample is not well mixed. As the materials are mixed for a longer period of time, more crosslinking appears to take place, and the samples become much firmer and more cohesive.

❸ *From your understanding of crosslinked polymers, discuss changes in properties, such as changes in elasticity, that you observed as you kneaded the samples for 8 minutes.*

The more the sample is mixed, the firmer, less sticky, and more elastic it becomes. The properties of this product change dramatically with the amount of mixing.

Section 5

Design a New Polymer Toy

● Student Research Activity

Many toys on the market are based on crosslinked polymers, ranging from Slime® to Super Balls®. The properties of crosslinked polymers may be changed by changing the nature of the polymer or by varying the ratio of the crosslinker to the polymer. Many products contain a mixture of polymers in order to achieve the desired product characteristics. Additives may also be used to modify the product characteristics.

In this research activity you will design a new polymer toy with characteristics you define, including desired elasticity, flow characteristics, feel, and other properties you wish to include.

Procedure

❶ Investigate the polymer toys available in the laboratory. Consider the following characteristics and record your observations below:

- elasticity—How well does the polymer bounce when formed into a ball?

- flow character—Does it stretch when pulled slowly? when pulled quickly? How far will it stretch?

- feel—Is it cool? wet? sticky?

- any other specific characteristics such as color, odor, or sensitivity to light

____

____

____

❷ Determine what characteristics you would like your new polymer toy to have. Record them in the space below.

____

____

____

____

____

❸ Using the polymers and crosslinker you used in Student Activity 4, experiment with varying the ratio of polymer to crosslinker, using more than one polymer mixture, mixing the polymer(s) and crosslinker in different orders, and adding other materials to achieve the characteristics you want in your new toy. Additives could include powdered chalk, plaster of paris, talcum powder, moisturizing lotion, or tiny Styrofoam® beads. As part of your research, keep detailed notes on your experiments and results. Prepare a series of batches of polymer, adjusting the mixture systematically to achieve optimum characteristics.

❹ Write a final research report including a complete procedure for preparing the final product (your new polymer toy); a detailed description of its characteristics, including research data; and a sample of the new polymer toy.

The purpose of this activity is to allow students to apply their knowledge of crosslinked polymers and mixing to create a new polymer toy. This activity effectively turns the classroom into a research laboratory. Students typically become very enthusiastic and proprietary about their own formulations.

Materials Notes

Using small, disposable drinking cups and wooden craft sticks for stirring minimizes cleanup time.

Getting Ready

Prepare 4% borax and 4% polyvinyl alcohol solutions (see the Instructor Notes for Student Activity 4). Collect a series of commercial polymer toys for the students to investigate: Slime®, Silly Putty®, Gak®, and Floam® are examples of commercial polymer toys that may be used to spark students' imagination.

Procedure Notes

Students may choose to try to duplicate the properties of one of the commercial products or to maximize some characteristic of the polymers that they have noted previously, such as bounciness or stretchiness.

Students should prepare a series of batches of their polymer, adjusting the mixture systematically to achieve optimum characteristics. Remind them to keep detailed records of their experimental data. The addition of chalk, plaster of paris, or talcum powder causes the formation of a semisolid compounded polymer with new properties. Solid additives create additional crosslink sites and act as fillers.

Made in the USA
Monee, IL
07 July 2026

56552021R00046